River

my guide to earth's habitats

Susan H. Gray

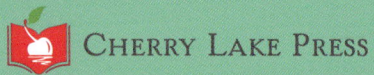

Published in the United States of America by Cherry Lake Publishing Group
Ann Arbor, Michigan
www.cherrylakepublishing.com

Reading Adviser: Beth Walker Gambro, MS, Ed., Reading Consultant, Yorkville, IL
Book Design: Jennifer Wahi
Illustrator: Jeff Bane

Photo Credits: © NotionPic/Shuttershock.com 2, 3, 24; © Mike Pellinni/Shuttershock.com, 5; © Damsea/Shuttershock.com, 7; © Xiao Zhou/Shuttershock.com, 9; © Augenstern/Shuttershock.com, 11; © Vladimir Arndt/Shuttershock.com, 13; © Ernie Cooper/Shuttershock.com, 15; © ABC photographs/Shuttershock.com, 17; © Bargais/Shuttershock.com, 19; © Joseph Scott Photography/Shuttershock.com, 21; © Volodymyr Burdiak/Shuttershock.com, 23; Cover, 6, 14, 22, Jeff Bane

Copyright © 2023 by Cherry Lake Publishing Group
All rights reserved. No part of this book may be reproduced or utilized in any form or by any means without written permission from the publisher.

Cherry Lake Press is an imprint of Cherry Lake Publishing Group.

Library of Congress Cataloging-in-Publication Data

Names: Gray, Susan Heinrichs, author. | Bane, Jeff, 1957- illustrator.
Title: River / by Susan H. Gray ; illustrated by Jeff Bane.
Description: Ann Arbor, Michigan : Cherry Lake Publishing, 2022. | Series: My guide to earth's habitats | Includes index. | Audience: Grades K-1
Identifiers: LCCN 2022005311 | ISBN 9781668909003 (hardcover) | ISBN 9781668910603 (paperback) | ISBN 9781668912195 (ebook) | ISBN 9781668913789 (pdf)
Subjects: LCSH: Stream ecology--Juvenile literature. | Rivers--Juvenile literature.
Classification: LCC QH541.5.S7 G73 2022 | DDC 577.6/4--dc23/eng/20220214
LC record available at https://lccn.loc.gov/2022005311

Printed in the United States of America
Corporate Graphics

table of contents

On the Move 4

Glossary 24

Index 24

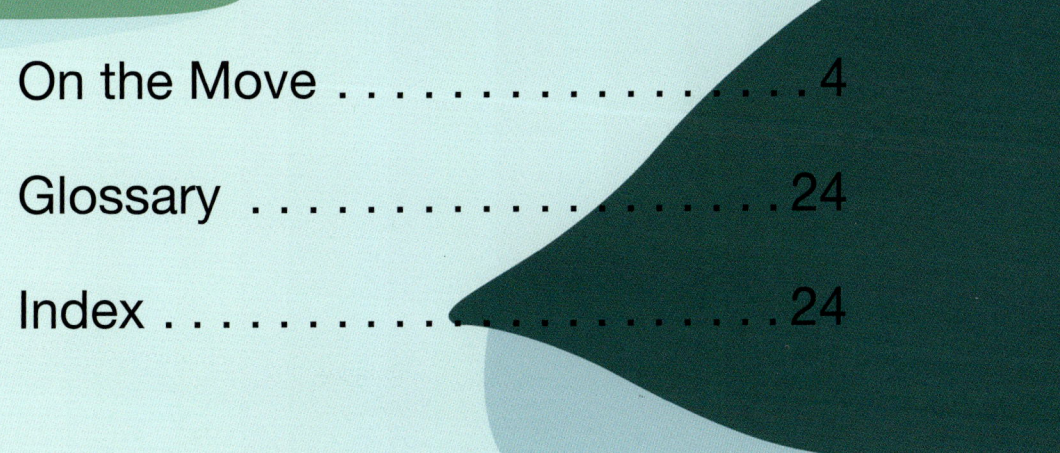

About the author: Susan H. Gray has a master's degree in zoology. She loves writing science books, especially about animals. Susan lives in Arkansas with her husband, Michael. They enjoy the state's many rivers.

About the illustrator: Jeff Bane and his two business partners own a studio along the American River in Folsom, California, home of the 1849 Gold Rush. When Jeff's not sketching or illustrating for clients, he's either swimming or kayaking in the river to relax.

On the Move

Rivers are always moving. Some are wild and fast. Others flow slowly.

A river bottom may be sandy. It could be rocky. Leaves and twigs might cover it.

Every river is **unique**. Each one has its own plants and animals.

Cattails grow in slow rivers. They live at the river's edge. Water is **shallow** there.

Waterweeds grow from sandy river bottoms. They have many green leaves.

Plants provide **shelter**. Baby fish live among them. Insects also hide on plants.

Mussels live on the river's bottom. **Gravel** helps them. They hang onto it. Then they don't wash away.

Fish are in rivers, too. Many feed on insects.

19

Herons live nearby. These birds stand in the water. They snap up fish that swim by.

21

Many living things **rely** on the river. They find food and safety there.

glossary & index

glossary

cattails (KAT-taylz) tall plants with brown, fuzzy flowers

gravel (GRAH-vuhl) small, loose pieces of rock

herons (HEHR-uhnz) water birds with long legs, necks, and bills

mussels (MUH-suhlz) water animals with soft bodies and two-part shells

rely (rih-LY) to depend on

shallow (SHAH-loh) not deep

shelter (SHEL-tuhr) a safe, protected place

unique (yoo-NEEK) unlike anything else

waterweeds (WAH-tuhr-weedz) types of plants that grow in water

index

bottom, 6, 16

fish, 14, 18, 20

flow, 4

grow, 10, 12

insects, 14, 18

sandy, 6, 12

unique, 8

water, 10, 20